# I'M SO *glad* I LIVE IN A WORLD WHERE THERE ARE *goats*

*Delci Plouffe*

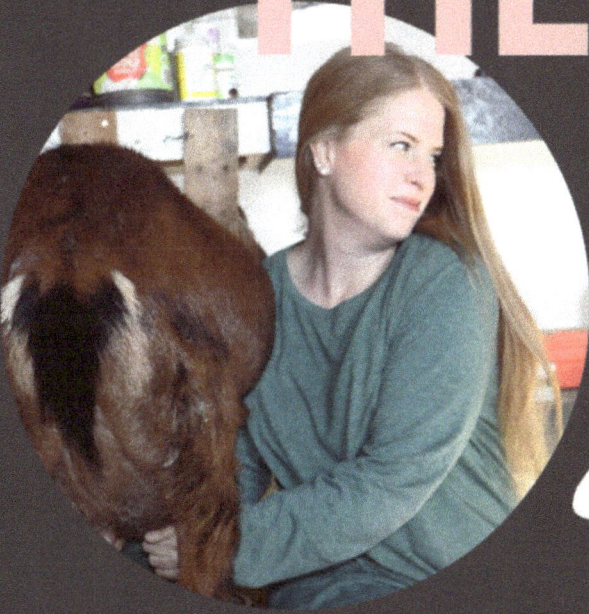

**My Parasite Control Plan Binder**
Copyright © 2022 by Delci Plouffe. All rights reserved.

ISBN  978-0-578-38274-6


A Life of Heritage Lewistown, MT 59457
www.alifeofheritage.com

Book design copyright © 2022 by Delci Plouffe. All rights reserved


*Refer to  **MY GOAT BINDER** for a complete record keeping system
vip.alifeofheritage.com/my-goat-binder-and-bonuses/

*I am NOT a vet. Find a qualified vet in your area. Many of the medications listed in this booklet are extra-label use but have been put together by vets and reputable breeders. Any goat can have a serious reaction at any time and many factors affect the outcomes, including what part of the country you live in. What works for some may not work for you and what works in your area may not work in other areas. Delci Plouffe or A Life of Heritage are not responsible for the outcome of any animal treated.*

# MY PARASITE CONTROL PLAN BINDER

# THE BIG PICTURE OF RAISING GOATS

Parasites are your biggest concern in raising goats.

Worms can cost you a great deal in the long run through...

- Treatment costs

- Your herd's performance lessening

- And death is a high possibility with high worm and parasite loads

And sadly, parasites have developed resistance to dewormers.

BUT the good news is that parasites can be managed!

## Your Goats...

- PREFER TO BROWSE

- ARE SELECTIVE EATERS

- LIKE TO ROAM

- CAN HANDLE SOME TOXINS OTHER LIVESTOCK CAN'T

- THEY LIKE VARIETY!

KNOWING YOUR GOAT'S PREFERENCES AND ALL THE INFORMATION IN THIS BOOKLET CAN HELP YOU KEEP YOUR GOATS HEALTHY AND FREE FROM HIGH WORM LOADS THAT CAN BE SO DETRIMENTAL TO THEIR HEALTH.

# WHAT CAUSES PARASITE PROBLEMS?

→ WE DO! ... HOW?

We cause parasite resistance by deworming them too often and incorrectly

We also don't keep them in good enough body condition year around

We force sheep and goats to eat too close to the ground

We create environments where parasites thrive

We also overcrowd them

And we "baby" our animals by keeping the animals we probably should cull

## INTERNAL PARASITE NUMBERS...

- Increase with the number of host animals (goats)
- Increase during warm, humid weather
- Increase when pastures are grazed too short

- Decrease during hot, dry weather
- Decrease if a non-host animal (cattle or horses) graze the same pasture
- Decrease with pasture rest time, as the larvae naturally die off

We can learn to manage them properly!

DEW DROP SHOWS LARVAE SUSPENDED IN THE MOISTURE. WHEN AN ANIMAL GRAZES, IT TAKES IN A LARGE AMOUNT OF PARASITES, WHICH CAN LEAD TO SICKNESS VERY QUICKLY.

SICK GOAT: EMACIATED, ROUGH HAIR COAT, LETHARGIC, BOTTLE JAW. THIS GOAT IS CONTAMINATING THE FARM THROUGH ITS DROPPINGS AND WILL BECOME A VERY EXPENSIVE ANIMAL IN THE LONG RUN.

# PRIMARY PARASITES

- **Barberpole worm (haemonchus contortus)** Most common worm (esp in the S.E. U.S.) that causes the most deaths. Dominates in summer months.
- **Brown stomach worm (telodorsagia-ostertagia-circumcincta)** More problematic in the fall and winter months.
- **Bankrupt worm (trichostrongylus colubriformis)** More problematic in the fall and winter months.
- **Coccidia (Eimera sp.)** Disease of stress and filth.
- **Tapeworm** Causes poor performance and may give potbellied appearance and give young animals more problems than mature animals.

# PARASITISM

Parasites are practically inevitable (all goats have them) but you can manage them so that parasitism is not evident.

Young animals are the ones most affected but does are susceptible and affected during the last month of pregnancy and right before and after kidding.

# SIGNS OF PARASITISM

- loss of condition
- rough hair coat
- bottle jaw
- low energy
- scours, diarrhea
- pale mucous membranes--anemia
- death

# why we can't rely on drugs alone

## 1
Parasites are becoming resistant to drugs-- particularly in sheep and goats

## 2
Parasites are already becoming resistant to the newest *anthelmintic drugs

## 3
We are running out of drugs to kill worms!

## 4
Anthelmintics should not be overused because they increase resistance

## 5
It is very important to use other methods in conjunction with anthelmintics

*Anthelmintics: antiparasitic drugs that expel parasitic worms and other internal parasites from the body by either stunning or killing them and without causing significant damage to the host.
They may also be called vermifuges or vermicides.

## CAUSES OF RESISTANCE

- **FREQUENT DEWORMING**
  - NO DEWORMER IS 100% EFFECTIVE, 100% OF THE TIME
  - FREQUENT DEWORMING INCREASES THE RATE THAT RESISTANCE DEVELOPS
- **DEWORMING ALL ANIMALS, REGARDLESS OF NEED**
  - THIS INCREASES RATE OF RESISTANCE
  - COSTS MORE MONEY
- **UNDER DOSING LEAVES MORE OF THE STRONG WORMS**
- **DEWORMING AND THEN MOVING TO A CLEAN PASTURE DEPOSITS ONLY THE RESISTANT WORMS INTO THE NEW PASTURE**

UNTREATED WORMS ARE CALLED "REFUGIA." DON'T OVERLOOK THIS!

HAVING SOME WORMS IN REFUGIA (NOT TREATED) INSURES THAT A LEVEL OF GENES REMAIN SENSITIVE TO DEWORMERS.

A SURVIVING POPULATION OF UNTREATED WORMS DIMINISHES THE FREQUENCY OF RESISTANT GENES. THEREFORE, WHEN A DEWORMER IS REQUIRED, IT WILL BE EFFECTIVE BECAUSE THE WORMS WILL BE SUSCEPTIBLE TO TREATMENT.

WHEN FEWER NUMBERS OF ANIMALS RECEIVE TREATMENT, THE REFUGIA POPULATION REMAINS LARGE.

THE MORE REFUGIA, THE BETTER.

SUSTAINABLE TECHNIQUES, LIKE MENTIONED IN THIS BOOKLET, FIGHT DRUG RESISTANCE BY INCREASING REFUGIA.

# YOU can control parasites...

### MOST IMPORTANT: PASTURE MANAGEMENT

- Avoid over-grazing by continually monitoring forage height.
  - Research indicates that most larvae can only travel about 1 inch off the ground so grazing close to the ground increases parasite ingestion.
  - Larvae migrate no more than 12 inches from a manure pile. Livestock not forced to eat close to their own manure will consume fewer larvae. Providing areas where animals can browse (eat brush, small trees, etc.) and eat higher off of the ground helps to control parasite problems.
- **Use proper stocking rate per pasture.**
- **Overstocking causes more worm deposits and forces animals to graze too close to the ground.**
- **Utilize multi-species grazing (small ruminants, cows, horses) because they do not share the same parasites**
  - For example: Cattle do not share the same internal parasites as sheep and goats. Cattle consume sheep and goat parasite larvae, which helps "clean" the pasture for the small ruminants.
- **Haying pastures removes and exposes larvae to the sun.**
- **Sanitation is a key factor.**

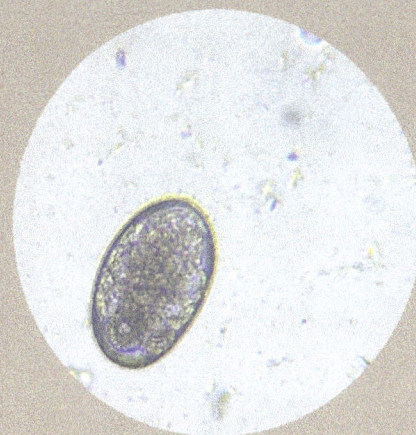

VIEW PARASITE MANAGEMENT FROM A HOLISTIC POINT OF VIEW AND DON'T RELY ON ONLY ONE METHOD PRESENTED IN THIS BOOKLET.

UTILIZE SEVERAL OF THESE TECHNIQUES AND ALWAYS KEEP A CLOSE EYE ON EACH GOAT FOR CHANGES IN BEHAVIOR AND BODY CONDITION.

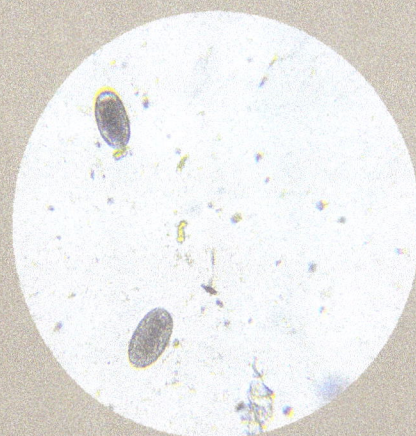

## ANIMAL SELECTION & MANAGEMENT

- Support optimal immune systems by providing good nutrition year around. Healthier animals have fewer problems.
- Select resistant animals by selecting resistant bucks and even consider selecting resistant breeds.
  - Spanish, Myotonic, Kiko. The Boar goat may not be as resistant.
- Cull animals needing more treatment and animals depositing a lot of eggs.

## GENERAL MANAGEMENT

- Keep feeders and water free of feces.
- Keep areas where animals congregate as clean as possible.
- Carefully choose, isolate and de-worm new animals.

## SMART DRENCHING

- Find out which dewormer works. Talk to your local vet and do a fecal egg count.
- Weigh animals prior to deworming (don't under dose!)
- Withhold feed 12-24 hours prior to dosing. Easy to do by feeding at night and giving wormer first thing in the morning and then feeding.
- Deliver dewormer over the tongue, in the back of the throat.
- Deworm only animals that need it!

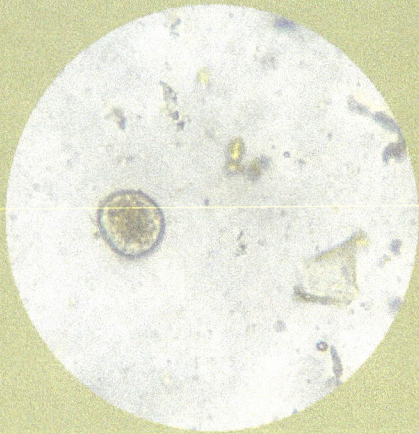

**RESEARCH SHOWS THAT 20 TO 30 PERCENT OF THE ANIMALS CARRY 70 TO 80 PERCENT OF THE WORMS.**

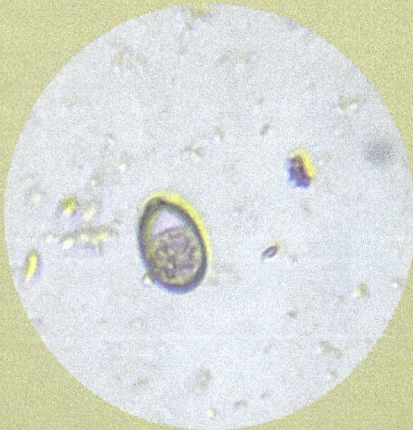

## FAMACHA

- Classifies animals based on level of anemia by examining the color of the eyelids.
- Only effective for haemonchus Barber Pole Worm (because it sucks blood).
- Slows dewormer resistance because fewer animals are treated.
- If you have a large herd/flock, examine 50, if 10-20% have 4's and 5's on the Famacha chart, you need to examine the whole herd.
- If they are in poor condition, treat the young, old, and about to kid.
- Total # of treatments can be decreased by up to 90% using this method.
- Keep records in "My Goat Binder"!
- Cull animals repeatedly treated.
- Treat less, save money!

## FOOD SELECTION

- Researchers have shown that big Trefoil, Sulla, Sanfoin, and sericea lespedeza (Chinese Bush Clover) are useful in controlling internal parasite infections. Providing condensed-tannin-containing forages is one way to boost the health of goats.
- Research has shown that Sericea is effective against internal parasites when grazed or when fed in dried forms, such as hay or pellets.

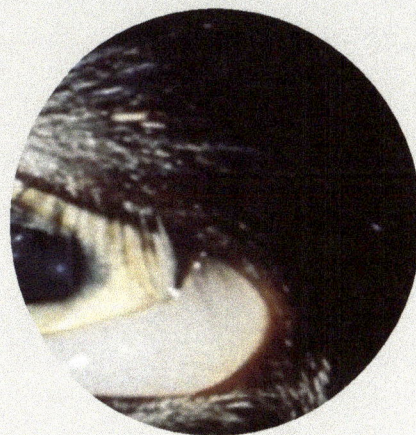

**REGULARLY CHECK YOUR GOATS EYELIDS FOR SIGNS OF ANEMIA AND HEAVY WORM LOADS.**

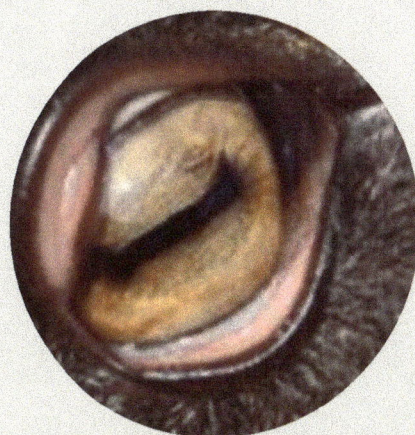

# COPPER BOLUSES

- Copper oxide wire particles have been proven to be an effective method of controlling the barber pole worm in goats.
- This method is only one piece of a holistic parasite management strategy and should not be the only method use for parasite management.
- Recommending copper bolus use for controlling worms does not endorse the use of high copper sulfate mineral mixes to control parasites.
- Be aware that sheep are highly susceptible to copper toxicity.
- Administer proper dosages using a pill gun or using the banana method. Some goats don't prefer bananas but may take copper rolled in a ball with peanut butter and rolled oats. You may try top dressing grain with copper rods or putting the rods in any number of treats your goats prefer.

*CUT HOLE IN BANANA PIECE*

*FILL WITH COPPER BOLUS*

*PUT BANANA TOP BACK ON*

*FEED TO GOAT*

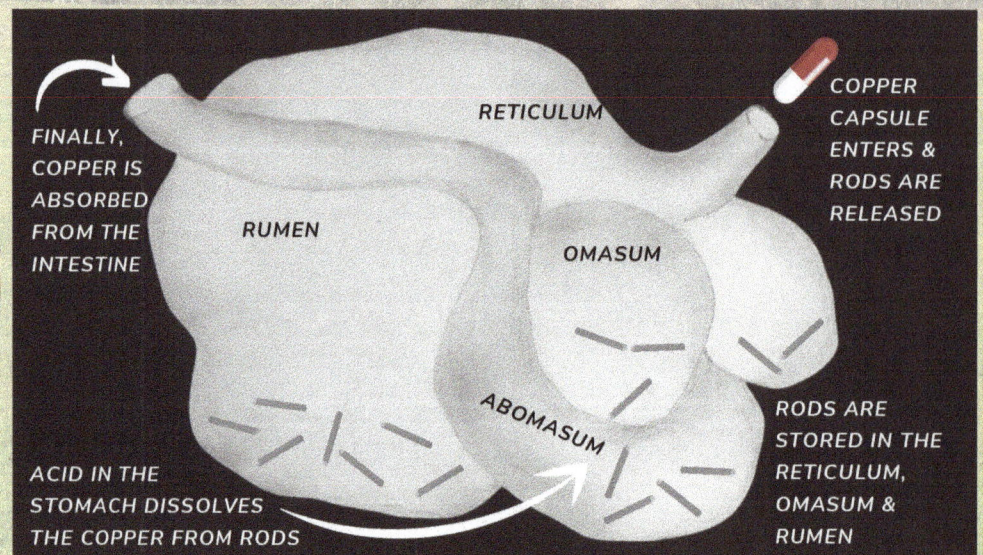

*COPPER CAPSULE ENTERS & RODS ARE RELEASED*

*RETICULUM*

*RUMEN*

*OMASUM*

*ABOMASUM*

*FINALLY, COPPER IS ABSORBED FROM THE INTESTINE*

*ACID IN THE STOMACH DISSOLVES THE COPPER FROM RODS*

*RODS ARE STORED IN THE RETICULUM, OMASUM & RUMEN*

**PRACTICE FECALS AND IDENTIFYING WORM EGGS UNTIL YOU ARE A MASTER AT KNOWING WHEN AND HOW TO WORM YOUR GOATS.**

## OTHER TECHNIQUES

- Garlic may repel worms and does boost a goat's immune system
  - Sick goat: give 1 clove garlic and 1 tsp peanut butter rolled in a ball
  - Garlic powder can be added to a mix of herbs that support their over all health.
- Nematode-trapping fungus (fungus traps parasite larva in the feces but may not be commercially available yet.)
- Vaccines are not available yet either.

## CONCLUSION

- Parasites are the biggest problem for goat producers.
- No technique is 100% effective.
- Several techniques should be used.
- Select for resistant animals.
- Do not over-treat (on a schedule or by worming all goats without a fecal check) because it causes resistance!

**Back**

Hair coat & body condition

- Brown Stomach Worm
- Bankrupt Worm
- Nodular Worm
- Barber Pole Worm
- Coccidia

**Eyelids**

Anemia

- Barber Pole Worm
- Liver Fluke
- Hook worms

**Jaw**

Edema
"Bottle Jaw"
Swelling

- Barber Pole Worm
- Liver Fluke
- Hook Worm

**Nose**

Discharge can indicate nose bots

- Nasal Botfly
- Lungworms
- Pneumonia (commonly affects goats weak from worms)

**Tail**

Loose stool, scours, soiled tail

- Brown Stomach Worm
- Bankrupt Worm
- Coccidia
- Nodlar Worm

# GOAT BODY CHECK

# PARASITE
# LIFECYCLE

Adult worms lay eggs in the goat which then pass to the pasture in feces

When the weather is suitable, the larvae hatch out

Larvae move from feces to grass through films of moisture

Larvae are eaten by goats

## ONE
Parasite larvae ingested

## SEVEN
Repeat cycle

## TWO
Adult larvae make residence in the body

## SIX
Animals ingest larvae

## THREE
Adults lay eggs

## FIVE
Eggs hatch & larvae crawl up blades of grass

## FOUR
Eggs passed in feces

*images not true to size

# Ruminant Internal Parasites

## Most Common Worms

### Barber Pole
Haemonchus

### Brown Stomach Worm
Ostertagia

### Bankrupt Worm
Trichostrongylus

### Coccidia
Protozoan: Causes Coccidiosis

### Tape Worm
Moniezia

### Small Intestinal Worm
Cooperia

### Whipworm
Trichuris

### Hookworm
Bunostomum

### Nodular Worm
Oesophagostomum

### Threadworm
Strongyloides

### Threadneck Worm
Nematodirus

### Mite Egg
Often mistaken for worm egg

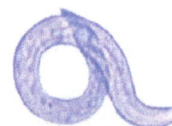

### Lungworm
Dictyocaulus

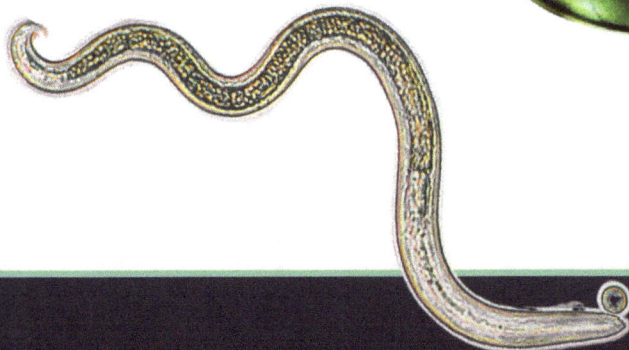

Please note that the color of the eggs and worms
will vary from microscope to microscope.

# FECAL EGG COUNT

## FIRST

- Collect poop.
- This can be done with a gloved finger in the rectum of the goat or immediately after the goat poops.
- Turn the glove or baggie inside out over the poop until used.
- Label with name of goat and date.
- If not using immediately, place in a bag, remove all air, seal well and label with date and goat's name. Then place in the fridge.
- Use your sample as soon as possible or keep in the fridge if you can't check right away.

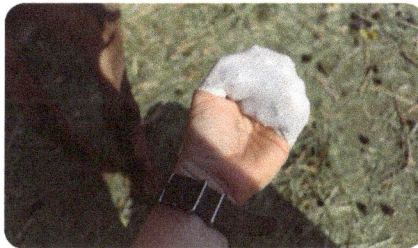

Eggs hatch within 12-24 hours if not refrigerated.

Analyze refrigerated feces within one week of collecting.

Check for: blood, odd color and/or mucus.

## SECOND

Put 2 grams (weigh on a scale, but usually 4-5 poop pills) in a container or cup, mash and then mix well with 15 ml (cc) of fecal flotation solution. Top off until there is 30 ml (cc) total of the mixture.

# FECAL EGG COUNT

**THIRD**

Shake the container/cup for 30 seconds to further break up the pills.

 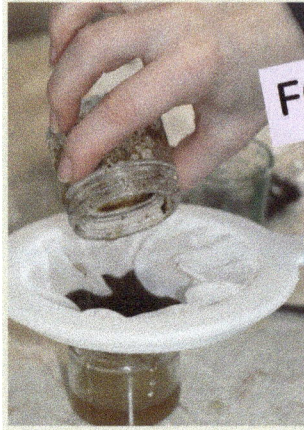

**FOURTH**

Strain the mixture through cheesecloth or fine strainer to strain out as much of the debris as possible.

Get the inside chamber of the McMaster slide wet, dry the outside of it. Then again rock the container/cup full of the mixture back and forth several times to make sure it is fully mixed.

**FIFTH**

Immediately and quickly, remove some solution with the eye dropper or syringe or pipette. Dispense the solution into both sides of the McMaster chamber, making sure the solution covers the entire area under the green grids. Then let the slide sit for 2-5 minutes, but no longer than an hour.

**SIXTH**

# FECAL EGG COUNT

Place the slide onto the microscope's stage and using the 10X, find one corner of the green grid and scan up and down the six lanes, counting all the worm eggs you see. For a closer look at an egg, change to 25X. Repeat this for the second grid on the slide.

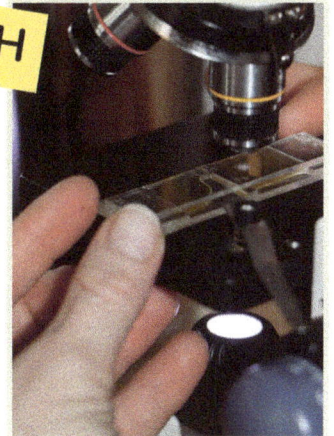

## Count these: oval, shell present, and cells that form larva

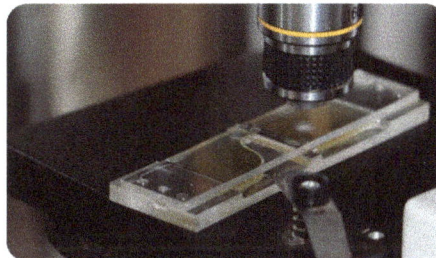

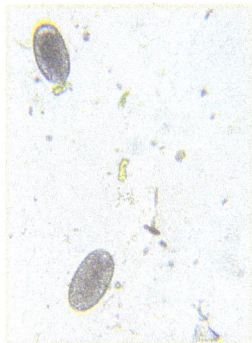

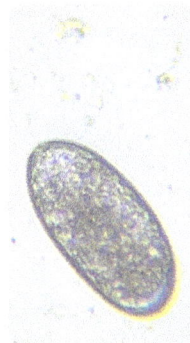

### EIGHTH

Multiply the number of worm eggs you see by 50 to get the Fecal Egg Count (FEC), i.e. eggs per gram of feces.

During this count you are counting the oval shaped strongyle/ trichostrongyle eggs & only making a note if you see strongyloides, tapeworm or coccidian eggs.

And count nothing outside of the chambers.

(First slide) 10 + 15  (Second slide) x 50 = 1250 epg

FEC: Fecal Egg Count
EPG: Eggs Per Gram

---

## PRACTICE IDENTIFYING EGGS

### AFTER STEP 4 ABOVE

You can also practice identifying eggs on a slide, apart from the McMaster slide.

Pour the liquid into a smaller topped container, making sure it "brims" over the top but doesn't spill. Then place a cover glass on top. Allow it to sit for 5 minutes. This allows the eggs to float to the top and stick to the cover glass.

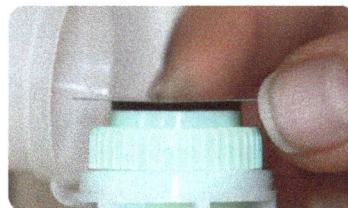

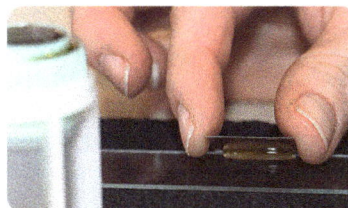

### Next

Place one side of the cover glass on the slide and let it fall down flat quickly in the center. Inspect under the microscope.

# FECAL EGG COUNT

**Limitations of this Method:**

Labs and people may do this procedure differently--establish how <u>you</u> do it and repeat exactly the same way each time.

Worms vary in egg producing capacity.

Strongyle eggs look the same and can not be differentiated at the egg state.

There is a day-to-day variation of the eggs distributed in the fecal matter.

Diarrhea and loose stools will most likely underestimate the fecal egg count.

Diarrhea *many times* means a high worm load.

A FEC (Fecal Egg Count) below 500 isn't usually an issue, so deworming may not be needed.

If a FEC of more than 500 exists, or if many coccidia oocysts are in your fecal sample, take appropriate corrective measures by medicating the goat properly.

Refer to the worming protocol in this book on worming properly.

Never tell someone, "Well, I wormed my goat so that can't be the problem..."

Treat your goat according to this book's worming procedure, and then repeat this Fecal Egg Count procedure 14 days after treatments to see if it was successful.

**Never assume that the treatments worked until you see the <u>results</u>.**

If you live in a worm problem area, do this randomly and regularly about once a month.

Then treat as necessary according to the findings in these tests.

Record all information and treatments in *<u>My Goat Binder</u>*.

# WORMER OPTIONS & DOSAGE

| CLASS | DRUG | COMMON NAME | DOSAGE | PREGNANCY |
|---|---|---|---|---|
| **BENZIMIDAZOLES** | FENBENDAZOLE | PANCUR\SAFEGUARD | 1 CC PER 10 LBS | YES |
| | ALBENDAZOLE | VALBAZEN | 1 CC PER 10 LBS | NO |
| | OXFENDAZOLE | SYNANTHIC | INJECTABLE GIVEN ORALLY 1 CC PER 35 LBS | NO |
| **MACROLYTIC LACTONES** — AVERMECTINS | IVERMECTIN | IVOMEC | Injectable 1%: Give orally. 1 cc per 20-30 lbs. Ivomec plus: Give orally. 1 cc per 20-30 lbs. Ivomec 1.87 paste: 1 cc per 30-50 lbs. Ivomec .08 sheep drench: 6 cc per 25 lbs. Mites: Inject 1 cc SQ per 30-50 lbs every wk for 3 wks. | YES |
| AVERMECTINS | DORAMECTIN | DECTOMAX | Mites: Inject 1 cc SQ per 30-50 lbs every wk for 3 wks. Injectable 1%: Give orally. 1 cc per 20-30 lbs. | YES |
| MILBEMYCINS | MOXIDECTIN | CYDECTIN | Oral Sheep Drench: 4.5 cc per 25 lbs Injectable: 1 cc orally per 25-50 lbs | YES |
| MILBEMYCINS | | QUEST | 1 cc per 100 lbs for normal worm load. 1 cc per 50 lbs for heavy worm load. | YES |
| AMINOQUINOLINE | PRAZIQUANTEL | EQUIMAX | 1 cc per 30-50 lbs. | YES |
| AMINOQUINOLINE | | ZIMECTERIN GOLD | 1 cc per 20-40 lbs. | YES |
| AMINOQUINOLINE | | QUEST PLUS | 1 cc per 50 lbs. | YES |

# WORMER OPTIONS & DOSAGE

| CLASS | DRUG | COMMON NAME | DOSAGE | PREGNANCY |
|---|---|---|---|---|
| **NICOTINIC AGONISTS** | | | | |
| IMIDAZOTHIAZOLES | LEVAMISOLE | PROHIBIT / LEVASOL / LEVAMED | Depends on how mixed. **DO NOT USE LEVAMISOLE UNLESS NO OTHER ANTHELMINTICS ARE WORKING! Repeat after 24 hours not 12 hours when treating for strongyles. | NO |
| TETRAHYDROPYRIMIDINES | PYRANTEL | STRONGID | 1 cc per 10 lbs. | YES |
| | RUMATEL | NEMATEL | NOT Recommended. | YES |

Directions:
When treating barber pole, bankrupt and other strongyloids, pick one wormer from two classes and give orally at the same time. Use two separate syringes. Repeat in 12 hours and repeat again in 10 days. If necessary, repeat again in 10 days.

Please note: These are guidelines to help you with your accuracy when treating your goats with anthelmintics. Please use under the guidance of a qualified vet. I am NOT a vet. Many of these are off-label use for goats. Many factors can affect the outcome and any goat can have serious reactions to medications at any time. Medications and anthelmintics may work for some and what works in some parts of the country may not work in another part. I am not responsible for the outcome.

# Effective Against...

| Brand name/Active Ingredient | Effective against |
|---|---|

**Albon (Sulfadimethoxine)**
**Baycox (Toltrazuril)**
**Corid (Amprolium)**
**Rumensin (Monensin)**
**Sulfaquinoxaline**

**PROTOZOA - COCCIDIA:**
- Coccidia (coccidiosis) Eimeria

**Prohibit (Levamisole)**

**ROUNDWORMS-NEMATODES:**
- Large stomach worm, barber pole worm, twisted worm
- Brown stomach worm
- Stomach/Intestinal hairworm, small stomach worm
- Thread-necked worm
- Hookworm
- Nodular worm
- Large-mouthed bowel worm
- Large lungworm
- Cooperia

**Cydectin (Moxidectin)**
**Dectomax (Doramectin)**
**Ivomec (Ivermectin)**
**Pyrantel Pamoate**
(Tetrahydropyrimidines)
**Quest (Moxidectin)**

**ROUNDWORMS-NEMATODES:**
- Large stomach worm, barber pole worm, twisted worm
- Brown stomach worm
- Stomach/Intestinal hairworm, small stomach worm
- Thread-necked worm
- Hookworm
- Nodular worm
- Large-mouthed bowel worm
- Whipworm
- Large lungworm
- Cooperia
- Strongyloides

# Effective Against...

| Brand name/Active Ingredient | Effective against |
|---|---|
| Panacur (Fenbendazole)<br>Safeguard (Fenbendazole) | **ROUNDWORMS-NEMATODES:**<br>• Large stomach worm, barber pole worm, twisted worm<br>• Brown stomach worm<br>• Stomach/Intestinal hairworm, small stomach worm<br>• Thread-necked worm<br>• Large-mouthed bowel worm<br>• Cooperia<br>• Strongyloides<br>**TAPEWORMS - CESTODES:**<br>• Broad Tapeworm |
| Clorsulon (in Ivomec Plus) | **FLUKES - TREMATODES:**<br>• Common liver fluke |
| Praziquantel (Biltricide) | **TAPEWORMS - CESTODES:**<br>• Broad Tapeworm |
| Valbazen (Albendazole) | **ROUNDWORMS-NEMATODES:**<br>• Large stomach worm, barber pole worm, twisted worm<br>• Brown stomach worm<br>• Stomach/Intestinal hairworm, small stomach worm<br>• Thread-necked worm<br>• Large-mouthed bowel worm<br>• Cooperia<br>**TAPEWORMS - CESTODES:**<br>• Broad Tapeworm<br>**FLUKES - TREMATODES:**<br>• Common liver fluke |
| IVOMEC & CYDECTIN | EXTERNAL PARASITES |

# more
# Treatment Information & Plans

## SpectoGard

- This is an antibiotic.

- It is not an anti-diarrheal, like Pepto, Kaopectate & Kaolin-Pectin.

- It treats Salmonella, E-Coli, and Bacterial enteritis.

- Dosage: 1 cc per 10 pounds.

- Give once or twice a day. And continue giving 3 days past normal poop.

- If there are no improvements after 48 hours, alternative treatments and diagnostics need to be utilized to figure out the real issue.

- Can be used for 3-5 days.

- If a goat still has diarrhea 24 hours after Coccidiosis treatment, this can be used.

## Coccidiosis

- Corid method:
  - Treat for 5 days and no less and at full strength.
  - Give it in a syringe to each kid according to weight. Do not use water treatment.
  - 6 cc/ 25 lbs,  12 cc/ 50 lbs,  18 cc/ 75 lbs,  24 cc/ 100 lbs
  - Treat with Fortified B Complex 100 mg/ml during or following this treatment by injecting 1 cc per 20 pounds SQ.

- Albon:
  - Albon 5% / Di-Methox -- 2.5 cc per 5 pounds orally 5 days.
  - Albon 12.5% / Di-Methox -- 1 cc per 5 pounds orally for 5 days.
  - Albon 40% -- Injectable given orally -- 1 cc per 15-20 pounds for 5 day.

# *more* Treatment Information & Plans

## Meningeal Worm

- 23 cc per 100 pounds of Fenbendazole (Safeguard or Panacur) orally for 5-7 days.

- Also give one single dose of Ivomec SQ.

- It is also beneficial to give banamine and dexamethasone. But dex can cause abortions in pregnant does.  Refer to dosages in "My Goat Binder".

## Levamisole (Prohibit)

- Only use this drug when all other anthelmintics aren't working. If you do use this, use a different anthelmintic the next time to avoid resistance of this powerful drug.  If resistance happens with this, nothing else will work after.

- Measure exactly (don't guess) and do not double dose.

- Dose on the back of the tongue (not just in the mouth).

- Large goats:
  - Mix with water:
    - 16 ounces with entire packet
    - 8 ounces with 1/2 packet
    - 4 ounces with 1/4 packet
    - Give 5.4 cc per 100 pounds. Repeat in 24 hours.

- Small goats:
  - Mix With Water:
    - 32 ounces with entire packet
    - 16 ounces with  1/2 packet
    - 8 ounces with 1/4 packet
    - Give 2.7 cc per 25 pounds or 10.8 cc per 100 pounds.

# The Most Common Worms

**Barberpole Worm-- Haemonchus Contortus**

**Brown Stomach Worm -- Osteargia Ostertagi**

**Banrupt Worm -- Trichostrongylus spp**

## Facts about these common worms:

- These 3 worms cause the most problems.
- You can not accurately ID these three worms by doing a fecal exam only. You will know that they are a species of strongyle but nothing more. You would have to hatch the eggs to identify them.
- But if diarrhea is present, you can make an educated guess that it's probably the Brown Stomach Worm, Bankrupt Worm or both causing the problems.
- If there are signs of anemia or bottle jaw, you can assume it is Barberpole Worm, which rarely causes diarrhea (unless there is a REALLY high count).
- Remember, all goats have some worms and coccidia even when they are healthy and they can even have a high count of coccidia without presenting any problems, issues or healthy concerns.

## How to treat:

- Choose 2 different wormers from 2 different classes.
- Give them at the same time from 2 different syringes.
- Then give again 12 hours later for a total of 2 treatments.
- Give again 10 days later. And then repeat in 10 days, if needed.

*Barberpole Worm Larvae can suck up to 800 cc or 27 oz of blood per day.

*Bleedout from worm dieout is a myth. Worm properly and timely to save your goats!

# Examples of Items in Fecal Checks

**Air Bubbles**

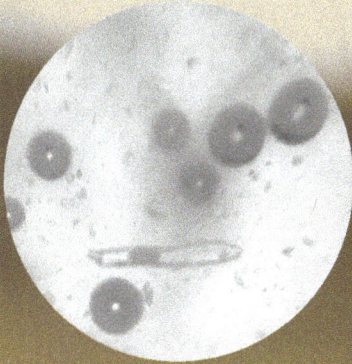

**Crystalized solution**

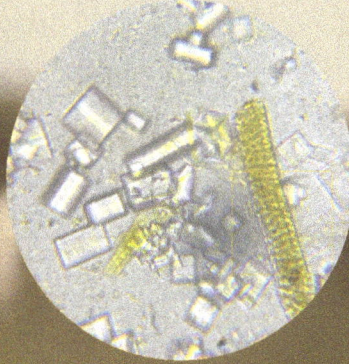

**Debris**

**Strongyle**

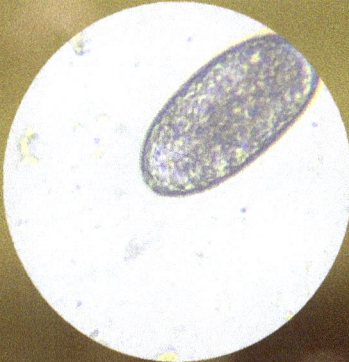

**Strongyle**

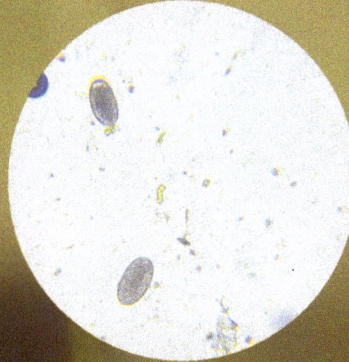

**Whipworm**

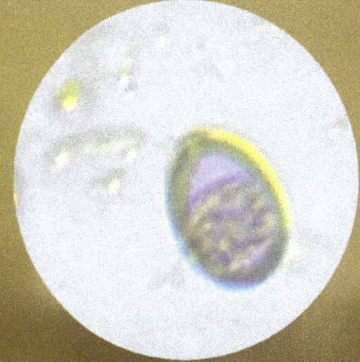

**Variety**

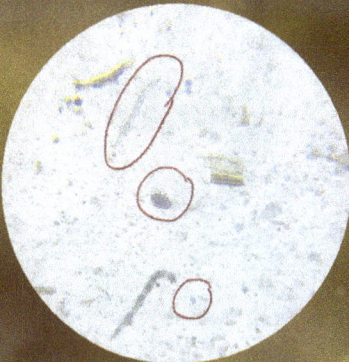

**Tapeworm**

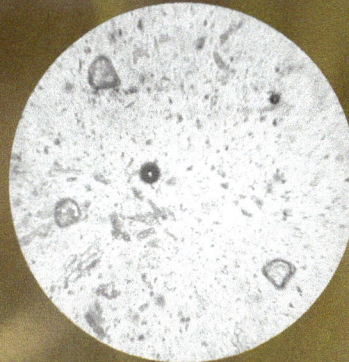

**Tapeworm**

# Examples of Items in Fecal Checks

**Coccidia**

**Coccidia**

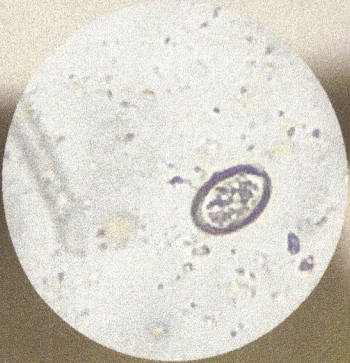

**Worm**

**Pollen**

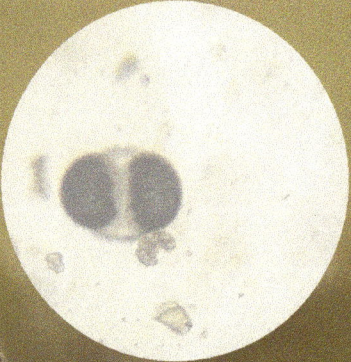

**Plant Hair**

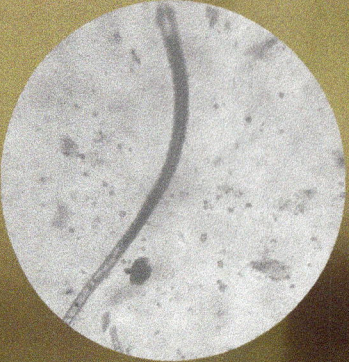

**Plant Hair**

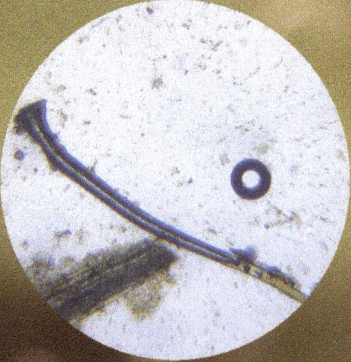

**Pollen**

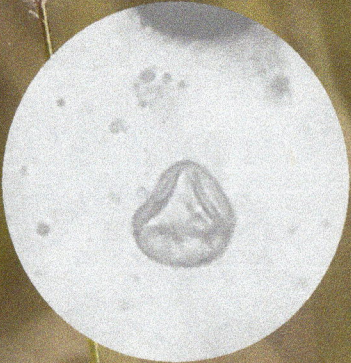

**Pollen**

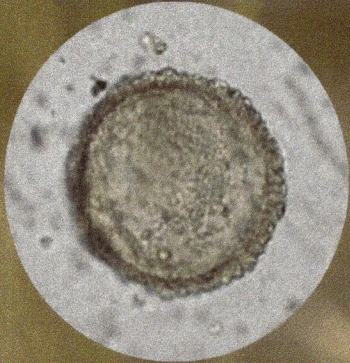

**Pollen**

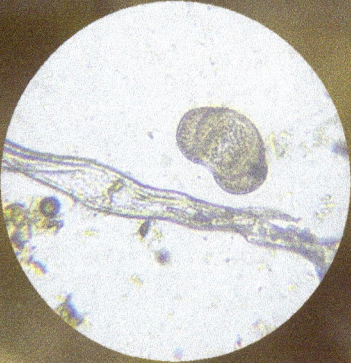

# MY Parasite
## CONTROL PLAN BINDER

www.ingramcontent.com/pod-product-compliance
Lightning Source LLC
Chambersburg PA
CBHW052346210326
41597CB00037B/6277